The Solar Energy Advantage

Maximizing Efficiency and Savings with Clean Power

Michelle Duncan

The Solar Energy Advantage

TABLE OF CONTENTS

Chapter 1: The Science Behind Solar Power

How Solar Panels Work

Solar panels, often seen glistening on rooftops or sprawling across fields, are the silent workhorses of renewable energy. They harness the sun's abundant energy, converting it into electricity that powers homes, businesses, and even entire communities. Understanding how these panels work involves delving into the fascinating world of photovoltaic technology, where science and innovation converge to create a sustainable energy solution.

At the heart of a solar panel lies the photovoltaic cell, a marvel of modern engineering. These cells are typically made from silicon, a semiconductor material that is abundant and efficient at converting sunlight into electricity. When sunlight strikes the surface of a photovoltaic cell, it excites electrons within the silicon. This excitation creates an electric field across the layers of the cell, causing electrons to flow and generate an electric current. This process, known as the photovoltaic effect, is the fundamental principle behind solar power generation.

The construction of a solar panel involves assembling multiple photovoltaic cells into a single unit. These cells are connected in series and parallel configurations to achieve the desired voltage and current output. The cells are then encapsulated in protective materials to shield them from environmental factors such as moisture and mechanical damage. A layer of tempered glass covers the cells, providing durability and allowing sunlight to pass through with minimal reflection. The entire assembly is framed in aluminum to provide structural support and facilitate mounting.

Solar panels come in various types, each with its own advantages and applications. Monocrystalline panels, known for their high efficiency and sleek appearance, are made from single-crystal silicon. They are ideal for installations where space is limited, as they produce more power per square meter compared to other types. Polycrystalline panels, on the other hand, are made from multiple silicon crystals and are generally more affordable, though slightly less efficient. Thin-film panels, made from materials like cadmium telluride or amorphous silicon, offer flexibility and lightweight properties, making them suitable for unconventional installations.

The role of inverters in solar systems is crucial, as they convert the direct current (DC) generated by solar panels into alternating current (AC), which is the standard form of electricity used in homes and businesses. Inverters also perform important functions such as maximizing power output through maximum power point tracking (MPPT) and ensuring the safety and stability of the electrical system. There are different types of inverters, including string inverters, microinverters, and power optimizers, each offering unique benefits depending on the specific requirements of the solar installation.

Innovations in solar technology continue to push the boundaries of efficiency and application. Bifacial solar panels, for instance, can capture sunlight from both sides, increasing energy yield by utilizing reflected light from surfaces like rooftops or the ground. Perovskite solar cells, a promising new technology, offer the potential for even higher efficiencies and lower production costs. Additionally, advancements in energy storage solutions, such as lithium-ion batteries, enable solar systems to store excess energy for use during periods of low sunlight, further enhancing their reliability and effectiveness.

The integration of smart technology into solar systems is another exciting development. Smart inverters and monitoring systems allow users to track their energy production and consumption in real-time, optimizing performance and identifying potential issues before they become significant problems. This level of control and insight empowers users to make informed decisions about their energy usage, contributing to greater efficiency and sustainability.

As solar technology continues to evolve, the potential applications for solar panels expand beyond traditional settings. Building-integrated photovoltaics (BIPV) incorporate solar cells directly into building materials, such as windows or facades, seamlessly blending energy generation with architectural design. Solar-powered transportation, from electric vehicles to solar planes, showcases the versatility and potential of harnessing solar energy in innovative ways.

The environmental benefits of solar panels are significant, as they provide a clean and renewable source of energy that reduces reliance on fossil fuels and decreases greenhouse gas emissions. By generating electricity from sunlight, solar panels contribute to a more sustainable and resilient energy future, mitigating the impacts of climate change and promoting energy independence.

Understanding how solar panels work is not only a matter of grasping the scientific principles behind them but also appreciating the broader impact they have on society and the environment. As technology advances and adoption increases, solar panels will continue to play a pivotal role in shaping a sustainable energy landscape, offering a viable solution to the world's growing energy needs.

Photovoltaic Technology Explained

Photovoltaic technology, a cornerstone of modern renewable energy solutions, is a testament to human ingenuity and the quest for sustainable power. At its core, this technology revolves around the conversion of sunlight into electricity, a process that is both elegant and efficient. The journey from sunlight to usable electrical power is a fascinating one, involving intricate scientific principles and cutting-edge engineering.

The foundation of photovoltaic technology lies in the photovoltaic cell, a device that captures sunlight and transforms it into electrical energy. These cells are primarily composed of semiconductor materials, with silicon being the most commonly used due to its abundance and favorable electronic properties. When photons from sunlight strike the surface of a photovoltaic cell, they impart energy to electrons within the semiconductor material. This energy boost allows the electrons to break free from their atomic bonds, creating electron-hole pairs. The movement of these free electrons generates an electric current, which can be harnessed for power.

Photovoltaic cells are typically organized into larger units known as solar panels or modules. These panels consist of multiple cells connected in series and parallel configurations to achieve the desired voltage and current output. The cells are encapsulated in protective materials to shield them from environmental factors such as moisture and mechanical stress. A layer of tempered glass covers the cells, allowing sunlight to pass through while providing durability and protection. The entire assembly is framed in aluminum, offering structural support and facilitating installation.

The efficiency of photovoltaic cells, or their ability to convert sunlight into electricity, is a critical factor in the performance of solar panels. Various types of photovoltaic cells have been developed to optimize efficiency and cost-effectiveness. Monocrystalline silicon cells, known for their high efficiency and uniform appearance, are made from single-crystal silicon. They are ideal for applications where space is limited, as they produce more power per square meter compared to other types. Polycrystalline silicon cells, on the other hand, are made from multiple silicon crystals and are generally more affordable, though slightly less efficient. Thin-film photovoltaic cells, made from materials such as cadmium telluride or amorphous silicon, offer flexibility and lightweight properties, making them suitable for unconventional installations.

Recent advancements in photovoltaic technology have led to the development of innovative cell designs and materials. Bifacial solar cells, for example, can capture sunlight from both sides, increasing energy yield by utilizing reflected light from surfaces like rooftops or the ground. Perovskite solar cells, a promising new technology, offer the potential for even higher efficiencies and lower production costs. These advancements are driving the evolution of photovoltaic technology, making it more accessible and efficient than ever before.

The integration of photovoltaic technology into various applications extends beyond traditional solar panels. Building-integrated photovoltaics (BIPV) incorporate solar cells directly into building materials, such as windows or facades, seamlessly blending energy generation with architectural design. This approach not only enhances the aesthetic appeal of buildings but also maximizes the use of available surface area for solar energy capture. Additionally, photovoltaic technology is being explored for use in transportation, with solar-powered vehicles and aircraft showcasing the versatility and potential of harnessing solar energy in innovative ways.

The environmental benefits of photovoltaic technology are significant, as it provides a clean and renewable source of energy that reduces reliance on fossil fuels and decreases greenhouse gas emissions. By generating electricity from sunlight, photovoltaic systems contribute to a more sustainable and resilient energy future, mitigating the impacts of climate change and promoting energy independence.

Understanding the intricacies of photovoltaic technology is not only a matter of grasping the scientific principles behind it but also appreciating the broader impact it has on society and the environment. As technology advances and adoption increases, photovoltaic systems will continue to play a pivotal role in shaping a sustainable energy landscape, offering a viable solution to the world's growing energy needs.

Types of Solar Panels: Monocrystalline, Polycrystalline, and Thin-Film

Solar panels have become a ubiquitous symbol of clean energy, yet not all panels are created equal. The diversity in solar panel technology allows for tailored solutions to meet specific energy needs and environmental conditions. Understanding the differences among monocrystalline, polycrystalline, and thin-film solar panels is crucial for making informed decisions about solar energy investments.

Monocrystalline solar panels are often recognized by their uniform, dark appearance and rounded edges. These panels are crafted from a single, continuous crystal structure, which is achieved through a process known as the Czochralski method. This method involves drawing a single crystal seed from molten silicon, resulting in a highly pure and efficient silicon wafer. The uniformity of the crystal structure allows for greater electron mobility, translating to higher efficiency rates. Monocrystalline panels typically boast efficiency ratings between 15% and 20%, making them one of the most efficient options available. This high efficiency is particularly advantageous in situations where space is limited, as they can generate more power per square meter compared to other types.

The manufacturing process of monocrystalline panels, while yielding high efficiency, is also more resource-intensive and costly. The precision required to produce single-crystal silicon results in higher production costs, which are often reflected in the price of the panels. However, the long-term energy yield and durability of monocrystalline panels can offset the initial investment, making them a popular choice for residential and commercial installations where efficiency and longevity are prioritized.

Polycrystalline solar panels, in contrast, are made from multiple silicon crystals melted together. This process is less complex and costly than the production of monocrystalline panels, resulting in a more affordable product. Polycrystalline panels are characterized by their bluish hue and square-shaped cells, which are cut from blocks of silicon. While they are generally less efficient than their monocrystalline counterparts, with efficiency ratings ranging from 13% to 16%, they offer a cost-effective solution for those looking to invest in solar energy without the higher upfront costs.

The lower efficiency of polycrystalline panels means they require more space to produce the same amount of energy as monocrystalline panels. This makes them suitable for installations where space is not a constraint, such as large commercial rooftops or ground-mounted solar farms. Additionally, advancements in polycrystalline technology have led to improvements in efficiency and performance, narrowing the gap between the two types of panels.

Thin-film solar panels represent a different approach to solar technology, utilizing a variety of materials beyond silicon. These panels are created by depositing one or more thin layers of photovoltaic material onto a substrate, such as glass, plastic, or metal. Common materials used in thin-film panels include cadmium telluride (CdTe), amorphous silicon (a-Si), and copper indium gallium selenide (CIGS). The manufacturing process for thin-film panels is less energy-intensive and allows for greater flexibility in design and application.

One of the most notable advantages of thin-film panels is their lightweight and flexible nature, which opens up a range of installation possibilities. They can be integrated into building materials, such as roofing shingles or facades, and are ideal for applications where traditional panels may not be feasible. However, thin-film panels generally have lower efficiency rates, typically between 10% and 12%, which means they require more surface area to generate the same amount of power as crystalline panels.

Despite their lower efficiency, thin-film panels perform well in low-light conditions and high temperatures, making them suitable for specific climates and applications. Their versatility and adaptability have made them a popular choice for innovative solar projects, such as portable solar chargers and solar-powered vehicles.

When choosing the right type of solar panel, several factors should be considered, including budget, available space, and specific energy needs. Monocrystalline panels offer high efficiency and durability, making them ideal for installations where space is limited and long-term performance is a priority. Polycrystalline panels provide a cost-effective solution for those with ample space and a focus on affordability. Thin-film panels, with their flexibility and adaptability, are well-suited for unique applications and environments.

The decision to invest in solar energy is a significant one, and understanding the nuances of different solar panel technologies can help ensure that the chosen solution aligns with both financial and environmental goals. As solar technology continues to evolve, the options available will expand, offering even more opportunities to harness the power of the sun in innovative and sustainable ways.

The Role of Inverters in Solar Systems

In the realm of solar energy systems, inverters play a pivotal role, acting as the bridge between the solar panels and the electrical grid or home appliances. While solar panels are responsible for capturing sunlight and converting it into direct current (DC) electricity, most homes and businesses operate on alternating current (AC) electricity. This is where inverters come into play, transforming the DC electricity generated by solar panels into AC electricity that can be used to power everyday devices.

The process of converting DC to AC is not merely a technical necessity; it is a sophisticated operation that involves several key functions. Inverters are equipped with advanced technology that ensures the efficient and safe conversion of electricity. One of the primary functions of an inverter is to perform maximum power point tracking (MPPT). This technology optimizes the power output from the solar panels by continuously adjusting the electrical operating point of the modules. By doing so, MPPT ensures that the solar panels are always operating at their maximum efficiency, regardless of changes in sunlight intensity or temperature.

Inverters also play a crucial role in maintaining the stability and safety of the solar energy system. They are responsible for synchronizing the frequency and voltage of the AC electricity with that of the grid, ensuring a seamless integration of solar power into the existing electrical infrastructure. This synchronization is vital for preventing disruptions and ensuring that the electricity generated by the solar panels can be effectively utilized or fed back into the grid.

There are several types of inverters available, each offering unique benefits and suited to different applications. String inverters are the most common type used in residential and commercial solar installations. In a string inverter system, multiple solar panels are connected in series to form a string, and the DC electricity from each string is fed into a single inverter. This setup is cost-effective and straightforward, making it a popular choice for many installations. However, string inverters can be less efficient in situations where panels are subject to shading or varying orientations, as the performance of the entire string is affected by the weakest panel.

Microinverters offer a solution to the limitations of string inverters by attaching a small inverter to each individual solar panel. This configuration allows each panel to operate independently, optimizing the power output of the entire system. Microinverters are particularly beneficial in installations where panels are subject to partial shading or different orientations, as they ensure that the performance of one panel does not impact the others. While microinverters tend to be more expensive upfront, they can offer higher overall efficiency and flexibility, making them an attractive option for complex installations.

Power optimizers are another technology that can enhance the performance of solar systems. Similar to microinverters, power optimizers are installed at the panel level, but they work in conjunction with a central inverter. Power optimizers condition the DC electricity from each panel, maximizing its efficiency before sending it to the central inverter for conversion to AC. This setup combines the benefits of both string inverters and microinverters, offering improved performance and cost-effectiveness.

In addition to converting electricity, modern inverters are equipped with smart technology that allows for monitoring and management of the solar energy system. Many inverters come with built-in communication capabilities, enabling users to track their energy production and consumption in real-time through online platforms or mobile apps. This level of insight empowers users to make informed decisions about their energy usage, optimize system performance, and identify potential issues before they become significant problems.

The integration of energy storage solutions with solar systems is another area where inverters play a crucial role. Hybrid inverters, designed to work with battery storage systems, manage the flow of electricity between the solar panels, batteries, and the grid. They ensure that excess solar energy is stored in batteries for later use, providing a reliable power supply even when the sun is not shining. This capability enhances the resilience and self-sufficiency of solar energy systems, making them an even more attractive option for those seeking energy independence.

As solar technology continues to advance, the role of inverters is becoming increasingly sophisticated. Innovations in inverter technology are driving improvements in efficiency, reliability, and functionality, making solar energy systems more accessible and effective than ever before. Understanding the role of inverters and the options available is essential for anyone considering a solar energy investment, as it ensures that the chosen system aligns with both current and future energy needs.

Innovations in Solar Technology

Solar technology has undergone remarkable transformations over the years, driven by the relentless pursuit of efficiency, sustainability, and accessibility. Innovations in this field are not only enhancing the performance of solar systems but also expanding their applications, making solar energy a more integral part of our daily lives. From advanced materials to cutting-edge designs, the landscape of solar technology is continually evolving, offering exciting possibilities for the future.

One of the most significant advancements in solar technology is the development of bifacial solar panels. Unlike traditional panels that capture sunlight only on one side, bifacial panels are designed to absorb light from both the front and back. This dual-sided capability allows them to harness additional energy from reflected sunlight, increasing their overall efficiency. Bifacial panels are particularly effective in environments with high albedo surfaces, such as snow-covered areas or reflective rooftops, where they can capture more diffuse light. This innovation not only boosts energy output but also maximizes the use of available space, making solar installations more productive.

Perovskite solar cells represent another groundbreaking advancement in solar technology. Named after the mineral with a similar crystal structure, perovskite materials have shown great promise in achieving high efficiency at a lower cost. These cells can be manufactured using simpler processes compared to traditional silicon-based cells, potentially reducing production costs significantly. Perovskite solar cells also offer flexibility in application, as they can be fabricated on lightweight and flexible substrates. This opens up new possibilities for integrating solar technology into a variety of surfaces, from building facades to wearable devices. The rapid progress in perovskite research suggests that these cells could soon rival or even surpass the efficiency of conventional solar technologies.

The integration of solar technology into building materials, known as building-integrated photovoltaics (BIPV), is transforming the way we think about energy generation. BIPV systems incorporate solar cells directly into construction materials, such as windows, roofs, and facades, seamlessly blending energy generation with architectural design. This approach not only enhances the aesthetic appeal of buildings but also maximizes the use of available surface area for solar energy capture. By turning buildings into power generators, BIPV systems contribute to energy efficiency and sustainability in urban environments, reducing reliance on external power sources and lowering carbon footprints.

Energy storage solutions are also playing a crucial role in the evolution of solar technology. The ability to store excess solar energy for use during periods of low sunlight is essential for ensuring a reliable and consistent power supply. Advances in battery technology, particularly lithium-ion and emerging solid-state batteries, are enhancing the capacity and efficiency of energy storage systems. These innovations enable solar systems to store more energy in a smaller space, providing greater flexibility and resilience. The integration of smart grid technology further enhances the management of energy storage, allowing for real-time monitoring and optimization of energy use.

The advent of smart solar technology is revolutionizing the way solar systems are monitored and managed. Smart inverters and monitoring systems provide users with detailed insights into their energy production and consumption, enabling them to optimize system performance and identify potential issues before they become significant problems. This level of control and transparency empowers users to make informed decisions about their energy usage, contributing to greater efficiency and sustainability. Additionally, smart technology facilitates the integration of solar systems with other renewable energy sources and smart home devices, creating a cohesive and intelligent energy ecosystem.

Solar technology is also making strides in the transportation sector, with solar-powered vehicles and infrastructure gaining traction. Solar panels integrated into electric vehicles can extend their range and reduce reliance on charging stations, while solar-powered charging stations offer a sustainable solution for recharging electric vehicles. These innovations are paving the way for a cleaner and more sustainable transportation future, reducing emissions and promoting energy independence.

The potential applications of solar technology extend beyond Earth, with space-based solar power systems being explored as a means of harnessing the sun's energy in space and transmitting it back to Earth. This ambitious concept involves deploying solar panels in orbit, where they can capture sunlight without atmospheric interference, and using microwave or laser technology to beam the energy back to ground-based receivers. While still in the experimental stage, space-based solar power holds the promise of providing a continuous and abundant source of clean energy, overcoming the limitations of terrestrial solar systems.

As solar technology continues to advance, the possibilities for harnessing the sun's energy are expanding in exciting and innovative ways. These advancements are not only enhancing the efficiency and accessibility of solar systems but also transforming the way we think about energy generation and consumption. By embracing these innovations, we can move towards a more sustainable and resilient energy future, where solar power plays a central role in meeting the world's growing energy needs.

Chapter 2: Assessing Solar Potential for Your Home

Evaluating Your Energy Needs

Understanding your energy needs is a crucial step in the journey toward adopting solar power. This process involves assessing your current energy consumption, anticipating future demands, and determining how solar energy can meet these needs effectively. By evaluating your energy requirements, you can make informed decisions about the size and type of solar system that will best suit your lifestyle and goals.

The first step in evaluating your energy needs is to conduct a thorough analysis of your current electricity usage. This involves reviewing your utility bills over the past year to identify patterns and fluctuations in your energy consumption. Pay attention to seasonal variations, as energy use often increases during extreme weather conditions when heating or cooling systems are in high demand. By understanding these patterns, you can gain insight into your baseline energy needs and identify opportunities for efficiency improvements.

To gain a more detailed understanding of your energy consumption, consider conducting an energy audit. This process involves examining your home or business to identify areas where energy is being used inefficiently. An energy audit can reveal hidden energy drains, such as outdated appliances, poor insulation, or inefficient lighting, that contribute to higher electricity bills. By addressing these issues, you can reduce your overall energy consumption and optimize the performance of your future solar system.

Once you have a clear picture of your current energy usage, it's important to consider any changes that may impact your future energy needs. This could include plans for home renovations, the addition of new appliances, or changes in household occupancy. For example, if you are planning to install an electric vehicle charging station or expand your living space, your energy needs may increase significantly. By anticipating these changes, you can ensure that your solar system is designed to accommodate future growth and avoid the need for costly upgrades down the line.

Another key factor in evaluating your energy needs is understanding the potential for energy savings through efficiency measures. Implementing energy-efficient practices and technologies can significantly reduce your overall energy consumption, allowing you to maximize the benefits of your solar system. Consider upgrading to energy-efficient appliances, installing LED lighting, and improving insulation to reduce heating and cooling demands. These measures not only lower your energy bills but also reduce the size and cost of the solar system required to meet your needs.

When evaluating your energy needs, it's also important to consider your energy goals and priorities. Are you primarily interested in reducing your carbon footprint, achieving energy independence, or saving money on your utility bills? Your goals will influence the design and configuration of your solar system, as well as any additional features you may choose to incorporate, such as battery storage or smart home technology. By aligning your solar investment with your energy goals, you can ensure that your system delivers the desired benefits and meets your expectations.

In addition to assessing your energy needs, it's essential to evaluate the potential for solar energy generation at your location. Factors such as roof orientation, shading, and local climate conditions can impact the efficiency and effectiveness of your solar system. Conducting a site assessment can help you determine the optimal placement and configuration of your solar panels to maximize energy production. Consider consulting with a solar professional to conduct a detailed site analysis and provide recommendations tailored to your specific circumstances.

Understanding your energy needs is a dynamic process that requires ongoing evaluation and adjustment. As your lifestyle and energy consumption patterns change, it's important to revisit your energy assessment and make any necessary modifications to your solar system. By staying informed and proactive, you can ensure that your solar investment continues to meet your needs and deliver value over the long term.

Evaluating your energy needs is a foundational step in the transition to solar power. By conducting a comprehensive analysis of your current and future energy requirements, you can make informed decisions about the size, type, and configuration of your solar system. This process not only ensures that your solar investment aligns with your goals and priorities but also maximizes the benefits of clean, renewable energy for your home or business.

Site Assessment: Roof Orientation and Shading

Conducting a thorough site assessment is a critical step in the successful installation of a solar energy system. Two of the most important factors to consider during this assessment are roof orientation and shading, as they significantly influence the efficiency and energy output of solar panels. By understanding these elements, you can optimize the placement and performance of your solar system, ensuring maximum energy generation and return on investment.

Roof orientation refers to the direction in which your roof faces, and it plays a crucial role in determining the amount of sunlight your solar panels will receive throughout the day. In the northern hemisphere, a south-facing roof is generally considered ideal for solar installations, as it receives the most direct sunlight over the course of the day. This orientation allows solar panels to capture the maximum amount of solar energy, leading to higher efficiency and energy production. However, east- and west-facing roofs can also be viable options, especially if they receive unobstructed sunlight during peak hours. While these orientations may result in slightly lower energy output compared to a south-facing roof, they can still provide substantial energy savings and environmental benefits.

When assessing roof orientation, it's important to consider the tilt or angle of your roof as well. The optimal tilt angle for solar panels depends on your geographic location and the angle of the sun's rays throughout the year. In general, the tilt angle should be close to the latitude of your location to maximize energy production. However, many solar installations are mounted flush with the roof to minimize costs and aesthetic impact, which may result in a suboptimal tilt angle. Despite this, modern solar panels are designed to perform efficiently across a range of angles, making them adaptable to various roof configurations.

Shading is another critical factor to consider during a site assessment, as it can significantly impact the performance of your solar system. Even partial shading on a single panel can reduce the energy output of the entire system, as it disrupts the flow of electricity through the interconnected panels. Common sources of shading include trees, chimneys, neighboring buildings, and other rooftop structures. To accurately assess shading, it's essential to conduct a shade analysis, which involves evaluating the potential sources of shade throughout the day and across different seasons.

A shade analysis can be performed using specialized tools and software that simulate the sun's path and identify areas of potential shading. This analysis helps determine the best placement for solar panels to minimize shading and maximize energy production. In some cases, strategic trimming or removal of trees and other obstructions may be necessary to optimize solar access. Additionally, advancements in solar technology, such as microinverters and power optimizers, can mitigate the effects of shading by allowing each panel to operate independently, ensuring that shaded panels do not impact the performance of the entire system.

Beyond roof orientation and shading, other site-specific factors should be considered during a site assessment. These include the structural integrity of the roof, available roof space, and local building codes and regulations. Ensuring that your roof is in good condition and capable of supporting the weight of solar panels is essential for a successful installation. It's also important to verify that your solar system complies with local zoning laws and building codes, which may dictate specific requirements for solar installations.

Conducting a comprehensive site assessment that takes into account roof orientation, shading, and other relevant factors is essential for optimizing the performance of your solar energy system. By carefully evaluating these elements, you can make informed decisions about the placement and configuration of your solar panels, ensuring that your system delivers maximum energy production and long-term benefits. This proactive approach not only enhances the efficiency and effectiveness of your solar investment but also contributes to a more sustainable and environmentally friendly energy future.

Calculating Solar Potential and Efficiency

Harnessing solar energy effectively begins with understanding the potential and efficiency of a solar installation. Calculating solar potential involves assessing how much sunlight a particular location receives and how this translates into usable energy. Efficiency, on the other hand, refers to how well a solar system converts sunlight into electricity. Both factors are crucial for designing a solar system that meets energy needs and maximizes return on investment.

The first step in calculating solar potential is to evaluate the solar insolation of a location. Solar insolation is the measure of solar radiation energy received on a given surface area during a specific time. It is typically expressed in kilowatt-hours per square meter per day ($kWh/m^2/day$). This measurement varies based on geographic location, time of year, and weather conditions. Regions closer to the equator generally receive higher solar insolation, making them more suitable for solar energy generation. However, even areas with lower insolation can benefit from solar power with the right system design.

To determine the solar potential of a site, it's essential to gather data on local solar insolation. This information can be obtained from various sources, including meteorological databases, solar maps, and online tools that provide solar radiation data for specific locations. By analyzing this data, you can estimate the average amount of sunlight your site receives and how it fluctuates throughout the year. This understanding is crucial for predicting the energy output of a solar system and planning for seasonal variations in energy production.

Once you have a clear picture of the solar potential, the next step is to assess the efficiency of the solar panels you plan to use. Solar panel efficiency is the percentage of sunlight that a panel can convert into usable electricity. This efficiency is influenced by several factors, including the type of solar panel, the quality of materials used, and the technology employed. Monocrystalline panels, for example, are known for their high efficiency, often exceeding 20%, while polycrystalline panels typically offer slightly lower efficiency rates. Thin-film panels, although less efficient, provide flexibility and are suitable for specific applications.

To calculate the expected energy output of a solar system, you need to consider both the solar potential and the efficiency of the panels. This involves multiplying the solar insolation by the efficiency of the panels and the total surface area of the installation. The result is an estimate of the energy that the system can produce over a given period, usually expressed in kilowatt-hours (kWh). This calculation helps determine whether the system can meet your energy needs and how much it can reduce reliance on conventional energy sources.

It's important to note that several factors can affect the actual efficiency of a solar system, including shading, temperature, and system losses. Shading from trees, buildings, or other obstructions can significantly reduce the amount of sunlight reaching the panels, decreasing their efficiency. Temperature also plays a role, as solar panels tend to operate less efficiently at higher temperatures. System losses, such as those occurring in inverters and wiring, can further reduce the overall efficiency of the system. To account for these factors, it's advisable to apply a derating factor to the calculated energy output, providing a more realistic estimate of the system's performance.

In addition to calculating potential and efficiency, it's essential to consider the financial aspects of a solar installation. This includes evaluating the cost of the system, potential savings on energy bills, and available incentives or rebates. By comparing the initial investment with the long-term savings, you can assess the financial viability of the solar project and determine the payback period. This analysis helps ensure that the solar system not only meets energy needs but also provides a sound financial return.

Understanding solar potential and efficiency is a critical component of designing an effective solar energy system. By accurately assessing these factors, you can optimize the performance of your solar installation, ensuring that it delivers maximum energy production and financial benefits. This knowledge empowers you to make informed decisions about your solar investment, contributing to a more sustainable and energy-efficient future.

Understanding Local Climate and Weather Patterns

Grasping the intricacies of local climate and weather patterns is essential for anyone considering the installation of a solar energy system. These factors play a significant role in determining the efficiency and effectiveness of solar panels, influencing both the amount of sunlight available and the overall energy output. By understanding these patterns, you can make informed decisions about the design and placement of your solar system, ensuring it meets your energy needs and maximizes its potential.

Climate refers to the long-term patterns of temperature, humidity, wind, and precipitation in a particular region. It provides a general overview of the environmental conditions that can be expected over extended periods. Weather, on the other hand, describes the short-term atmospheric conditions that can change from day to day. Both climate and weather have a direct impact on solar energy generation, as they determine the amount of sunlight that reaches the Earth's surface.

One of the primary climate factors to consider is the average solar insolation in your area. Solar insolation measures the amount of solar radiation received per unit area and is typically expressed in kilowatt-hours per square meter per day (kWh/m²/day). Regions with high solar insolation, such as deserts or areas near the equator, are ideal for solar energy generation, as they receive abundant sunlight throughout the year. However, even regions with lower insolation can benefit from solar power, provided the system is designed to account for local conditions.

Temperature is another critical climate factor that affects solar panel performance. Solar panels are generally more efficient at converting sunlight into electricity at lower temperatures. High temperatures can reduce the efficiency of solar cells, leading to decreased energy output. This phenomenon, known as the temperature coefficient, varies among different types of solar panels. When selecting panels for your solar system, it's important to consider their temperature coefficient and how it may impact performance in your local climate.

Precipitation and cloud cover are also important considerations, as they can significantly affect the amount of sunlight reaching your solar panels. Regions with frequent rain or heavy cloud cover may experience reduced solar energy generation, particularly during certain seasons. However, advancements in solar technology have improved the ability of panels to generate electricity even under diffuse light conditions. Additionally, rain can have a beneficial effect by naturally cleaning the panels, removing dust and debris that can obstruct sunlight.

Wind patterns can influence the design and installation of solar systems, particularly in areas prone to strong winds or storms. While wind itself does not directly impact solar energy generation, it can affect the structural integrity of solar installations. Ensuring that your solar panels are securely mounted and capable of withstanding local wind conditions is essential for maintaining their performance and longevity.

Seasonal variations in climate and weather patterns should also be taken into account when planning a solar installation. In many regions, solar energy generation is higher during the summer months when days are longer and sunlight is more intense. Conversely, energy production may decrease during the winter months due to shorter days and lower sun angles. By understanding these seasonal fluctuations, you can anticipate changes in energy output and plan accordingly, such as by incorporating energy storage solutions to balance supply and demand.

Local microclimates can also play a role in solar energy generation. Microclimates are small-scale climate variations that occur within a larger climate zone, often influenced by geographical features such as mountains, bodies of water, or urban environments. These microclimates can create unique conditions that affect solar potential, such as increased fog in coastal areas or higher temperatures in urban heat islands. Conducting a site-specific assessment can help identify any microclimatic factors that may impact your solar system's performance.

Understanding local climate and weather patterns is a vital component of designing an effective solar energy system. By considering factors such as solar insolation, temperature, precipitation, wind, and seasonal variations, you can optimize the placement and configuration of your solar panels to maximize energy production. This knowledge empowers you to make informed decisions about your solar investment, ensuring that it delivers reliable and sustainable energy for years to come.

Tools and Resources for Solar Assessment

Embarking on a solar energy project requires a comprehensive understanding of the tools and resources available for solar assessment. These tools are essential for evaluating the feasibility of solar installations, optimizing system design, and ensuring maximum energy production. By leveraging these resources, you can make informed decisions that align with your energy goals and financial considerations.

One of the most valuable tools for solar assessment is solar mapping software. These programs provide detailed information about solar potential by analyzing geographic data, weather patterns, and shading effects. They offer insights into the solar insolation of a specific location, helping you determine the best placement for solar panels. Popular solar mapping tools include PVGIS (Photovoltaic Geographical Information System), HelioScope, and Aurora Solar. These platforms allow users to simulate different scenarios, assess energy output, and optimize system design based on site-specific conditions.

In addition to mapping software, solar calculators are indispensable for estimating the energy production and financial viability of a solar project. These calculators take into account factors such as panel efficiency, system size, and local electricity rates to provide an estimate of potential savings and payback periods. By inputting relevant data, you can gain a clearer understanding of the economic benefits of solar energy and make informed decisions about your investment. Many solar calculators are available online, offering user-friendly interfaces and customizable options to suit individual needs.

For those seeking a more hands-on approach, solar assessment kits provide the tools necessary to conduct a detailed site evaluation. These kits typically include devices such as solar pathfinders, which help identify shading obstacles and determine optimal panel placement. They may also contain pyranometers, which measure solar radiation levels, and inclinometers, which assess roof tilt angles. By using these tools, you can gather precise data about your site and make informed decisions about system design and configuration.

Access to reliable data is crucial for accurate solar assessment, and several resources provide valuable information on solar potential and weather patterns. The National Renewable Energy Laboratory (NREL) offers a wealth of data and resources, including the Solar Resource Data website, which provides access to solar radiation data for locations worldwide. Similarly, the Global Solar Atlas, developed by the World Bank, offers high-resolution maps and data on solar potential, helping users identify suitable sites for solar installations.

Local government agencies and utility companies can also be valuable sources of information and support for solar assessment. Many offer programs and incentives to encourage the adoption of solar energy, providing resources such as rebates, tax credits, and grants. By reaching out to these organizations, you can gain access to valuable information about local regulations, permitting processes, and financial incentives that can enhance the feasibility of your solar project.

Networking with solar professionals and industry experts is another effective way to gain insights and guidance for solar assessment. Attending industry conferences, workshops, and webinars can provide opportunities to learn from experienced practitioners and stay informed about the latest trends and technologies in solar energy. Engaging with online forums and communities dedicated to solar energy can also offer valuable advice and support from individuals who have successfully navigated the solar assessment process.

Educational resources, such as books, online courses, and tutorials, can further enhance your understanding of solar assessment and system design. These resources cover a wide range of topics, from the basics of solar energy to advanced system optimization techniques. By investing time in learning and expanding your knowledge, you can build the confidence and expertise needed to make informed decisions about your solar project.

Utilizing a combination of tools and resources for solar assessment is essential for ensuring the success of your solar energy project. By leveraging mapping software, calculators, assessment kits, and reliable data sources, you can conduct a thorough evaluation of your site's solar potential and design a system that meets your energy needs. Engaging with industry experts and educational resources further enhances your understanding and empowers you to make informed decisions that maximize the benefits of solar energy. This comprehensive approach not only optimizes the performance of your solar installation but also contributes to a more sustainable and energy-efficient future.

Chapter 3: Financial Considerations and Incentives

Cost Analysis: Upfront Investment vs. Long-Term Savings

Navigating the financial landscape of solar energy involves a careful examination of the upfront investment and the long-term savings associated with solar installations. Understanding this balance is crucial for making informed decisions that align with both your budget and your energy goals. By conducting a thorough cost analysis, you can determine the financial viability of a solar project and ensure that it delivers the desired economic benefits over time.

The initial cost of a solar energy system is often the most significant barrier for potential adopters. This upfront investment includes the cost of solar panels, inverters, mounting hardware, and installation labor. Additionally, there may be expenses related to permits, inspections, and any necessary upgrades to your electrical system. While these costs can be substantial, it's important to consider them in the context of the long-term savings and benefits that solar energy provides.

One of the primary financial advantages of solar energy is the reduction in electricity bills. By generating your own electricity, you can significantly decrease or even eliminate your reliance on grid power, leading to substantial savings over the life of the system. The extent of these savings depends on several factors, including the size of the solar system, local electricity rates, and your energy consumption patterns. In regions with high electricity costs, the savings from solar energy can be particularly pronounced, resulting in a shorter payback period and greater return on investment.

To accurately assess the long-term savings of a solar system, it's essential to consider the system's expected lifespan and performance. Most solar panels come with warranties that guarantee performance for 25 years or more, providing a reliable source of energy for decades. During this time, the cost of electricity from the grid is likely to increase, further enhancing the financial benefits of solar energy. By factoring in these potential savings, you can gain a clearer understanding of the overall value of your solar investment.

In addition to reducing electricity bills, solar energy systems can increase the value of your property. Homes and businesses with solar installations are often more attractive to buyers, as they offer the promise of lower energy costs and a reduced carbon footprint. This increased property value can offset some of the initial investment, providing an additional financial incentive for adopting solar energy.

Government incentives and rebates play a crucial role in making solar energy more affordable and accessible. Many regions offer financial incentives to encourage the adoption of renewable energy, including tax credits, rebates, and grants. These incentives can significantly reduce the upfront cost of a solar system, making it more financially viable for a wider range of individuals and businesses. It's important to research and take advantage of any available incentives in your area to maximize the financial benefits of your solar investment.

Financing options are another important consideration when evaluating the cost of a solar system. Many solar providers offer financing plans that allow you to spread the cost of the system over several years, reducing the initial financial burden. Options such as solar loans, leases, and power purchase agreements (PPAs) provide flexibility and make solar energy more accessible to those who may not have the capital for an upfront purchase. By exploring these financing options, you can find a solution that fits your budget and financial goals.

While the financial benefits of solar energy are significant, it's important to consider the potential costs associated with maintenance and system performance. Solar panels require minimal maintenance, but periodic cleaning and inspections are necessary to ensure optimal performance. Additionally, inverters and other system components may need replacement over time, adding to the overall cost of ownership. By budgeting for these expenses, you can ensure that your solar system continues to deliver reliable energy and financial savings throughout its lifespan.

Conducting a comprehensive cost analysis that considers both the upfront investment and long-term savings is essential for making informed decisions about solar energy. By evaluating the financial benefits of reduced electricity bills, increased property value, and available incentives, you can determine the overall value of your solar investment. This analysis empowers you to make decisions that align with your financial goals and contribute to a more sustainable and energy-efficient future.

Government Incentives and Tax Credits

Navigating the landscape of government incentives and tax credits is a pivotal aspect of making solar energy more accessible and financially viable. These incentives are designed to encourage the adoption of renewable energy by reducing the initial costs associated with solar installations. Understanding the various programs available can significantly impact the overall affordability and attractiveness of solar energy for both residential and commercial users.

One of the most well-known incentives is the federal Investment Tax Credit (ITC) in the United States. This program allows homeowners and businesses to deduct a significant percentage of the cost of installing a solar energy system from their federal taxes. The ITC has been instrumental in driving the growth of solar energy across the country, making it a cornerstone of solar financing strategies. It's important to note that the percentage of the tax credit can vary over time, so staying informed about current rates and deadlines is crucial for maximizing benefits.

In addition to federal incentives, many states offer their own programs to promote solar energy adoption. These state-level incentives can take various forms, including tax credits, rebates, and grants. For example, some states provide additional tax credits that can be combined with the federal ITC, further reducing the net cost of a solar installation. Others offer direct rebates that provide immediate financial relief upon installation. Researching the specific incentives available in your state is essential for understanding the full range of financial benefits you can access.

Net metering is another valuable incentive that can enhance the financial viability of solar energy systems. This program allows solar system owners to receive credit for excess electricity they generate and feed back into the grid. Essentially, your utility meter runs backward when your system produces more energy than you consume, resulting in lower electricity bills. Net metering policies vary by state and utility company, so it's important to understand the specific terms and conditions that apply to your location.

Property tax exemptions are also a common incentive offered by many states and municipalities. These exemptions prevent the increase in property taxes that would typically result from the added value of a solar energy system. By excluding the value of the solar installation from property tax assessments, these exemptions help maintain affordability and encourage more homeowners to invest in solar energy.

Sales tax exemptions can further reduce the cost of solar installations by eliminating the sales tax on solar equipment purchases. This incentive is particularly beneficial in states with high sales tax rates, as it can lead to substantial savings on the overall cost of a solar system. As with other incentives, the availability and specifics of sales tax exemptions vary by state, so it's important to verify the details for your location.

For commercial solar projects, additional incentives may be available, such as accelerated depreciation through the Modified Accelerated Cost Recovery System (MACRS). This program allows businesses to recover the cost of solar investments more quickly by depreciating the value of the system over a shorter period. By reducing taxable income, MACRS can significantly enhance the financial return on commercial solar projects.

Beyond financial incentives, some states and local governments offer technical assistance and support programs to facilitate the adoption of solar energy. These programs may include resources for navigating permitting processes, guidance on system design and installation, and access to educational materials. By providing this support, governments aim to reduce barriers to entry and streamline the transition to solar energy.

It's important to recognize that government incentives and tax credits are subject to change, as they are often influenced by policy decisions and budgetary considerations. Staying informed about current programs and any upcoming changes is essential for maximizing the financial benefits of solar energy. Engaging with local solar installers and industry associations can provide valuable insights and updates on the latest developments in solar incentives.

Understanding and leveraging government incentives and tax credits is a critical component of making solar energy more affordable and accessible. By taking advantage of these programs, you can significantly reduce the upfront costs of solar installations and enhance the long-term financial benefits of your investment. This strategic approach not only supports the transition to renewable energy but also contributes to a more sustainable and economically viable future.

Financing Options: Loans, Leases, and Power Purchase Agreements

Exploring the financial pathways to solar energy adoption reveals a variety of options that cater to different needs and circumstances. Understanding these financing mechanisms— loans, leases, and power purchase agreements (PPAs)—is crucial for making solar energy accessible and affordable. Each option offers unique benefits and considerations, allowing individuals and businesses to choose the best fit for their financial situation and energy goals.

Loans are a popular choice for those who wish to own their solar energy system outright. By securing a loan, you can finance the upfront cost of the solar installation and repay it over time, similar to a mortgage. This approach allows you to benefit from any available tax credits and incentives, as you retain ownership of the system. Solar loans can be obtained from various sources, including banks, credit unions, and specialized solar financing companies. Interest rates and terms vary, so it's important to shop around and compare offers to find the most favorable conditions.

One of the key advantages of solar loans is the potential for long-term savings. By owning the system, you can significantly reduce or eliminate your electricity bills, leading to substantial savings over the life of the system. Additionally, once the loan is paid off, you continue to benefit from free electricity generation, maximizing your return on investment. However, it's important to consider the monthly loan payments and ensure they fit within your budget, as they will offset some of the immediate savings on your energy bills.

Leases offer an alternative financing option for those who prefer not to own the solar system. Under a solar lease, a third-party company installs and maintains the solar panels on your property, and you pay a fixed monthly fee for the use of the system. This arrangement allows you to access solar energy without the upfront costs and responsibilities of ownership. Leases typically include maintenance and monitoring services, providing peace of mind and ensuring optimal system performance.

The primary benefit of a solar lease is the ability to enjoy the benefits of solar energy with minimal financial risk. Since the leasing company owns the system, they are responsible for any repairs or maintenance, reducing the burden on the homeowner. Additionally, leases often come with predictable monthly payments, making it easier to budget for energy costs. However, it's important to carefully review the terms of the lease agreement, as some contracts may include annual escalators that increase the monthly payment over time.

Power Purchase Agreements (PPAs) are another financing option that allows you to access solar energy without owning the system. Similar to a lease, a third-party company installs and maintains the solar panels on your property. However, instead of paying a fixed monthly fee, you agree to purchase the electricity generated by the system at a predetermined rate. This rate is typically lower than the local utility rate, providing immediate savings on your energy bills.

PPAs offer the advantage of no upfront costs and reduced energy expenses, making them an attractive option for those looking to lower their electricity bills without a significant financial commitment. The PPA provider is responsible for the system's maintenance and performance, ensuring that it operates efficiently throughout the contract term. As with leases, it's important to review the terms of the PPA carefully, as some agreements may include price escalators or other conditions that affect long-term savings.

When considering financing options, it's essential to evaluate your financial goals, risk tolerance, and long-term plans. Ownership through a loan may offer the greatest financial benefits over time, but it requires a higher initial investment and assumes responsibility for system maintenance. Leases and PPAs provide a more hands-off approach with lower upfront costs, but they may offer less long-term savings compared to ownership.

It's also important to consider the impact of financing options on property value and potential resale. Owning a solar system can increase the value of your property, as it offers the promise of reduced energy costs for future buyers. Leases and PPAs, on the other hand, may require transferring the agreement to the new owner, which could complicate the sale process. Understanding these implications can help you make an informed decision that aligns with your long-term plans.

Engaging with solar financing experts and conducting thorough research can provide valuable insights into the best financing option for your situation. By exploring the various pathways to solar energy adoption, you can find a solution that meets your financial needs and supports your transition to renewable energy. This strategic approach not only enhances the affordability of solar energy but also contributes to a more sustainable and energy-efficient future.

Return on Investment: Calculating Payback Period

Calculating the return on investment (ROI) and payback period for solar energy systems is a crucial step for homeowners considering this sustainable energy source. Understanding these financial metrics helps in making informed decisions about the feasibility and benefits of installing solar panels. The ROI and payback period are influenced by several factors, including the initial cost of the system, energy savings, government incentives, and the lifespan of the solar panels.

The initial investment in a solar energy system can be substantial, encompassing the cost of solar panels, inverters, installation, and any additional equipment needed for optimal performance. However, this upfront cost is often offset by significant long-term savings on electricity bills. To calculate the ROI, one must first determine the total cost of the solar system and compare it to the expected savings over time. This involves estimating the amount of electricity the system will generate and the corresponding reduction in utility bills.

Energy savings are a primary component of the ROI calculation. By generating their own electricity, homeowners can reduce or even eliminate their reliance on the grid, leading to substantial savings. The amount of savings depends on several factors, including the size of the solar system, local electricity rates, and the amount of sunlight the location receives. In regions with high electricity costs and abundant sunshine, the savings can be particularly significant.

Government incentives and tax credits play a crucial role in enhancing the ROI of solar energy systems. Many governments offer financial incentives to encourage the adoption of renewable energy, such as tax credits, rebates, and grants. These incentives can significantly reduce the initial cost of the system, thereby improving the ROI. It's important for homeowners to research and understand the incentives available in their area, as they can vary widely depending on location and government policies.

The payback period is another important metric to consider when evaluating the financial viability of a solar energy system. It represents the time it takes for the savings generated by the system to equal the initial investment. A shorter payback period indicates a more attractive investment. To calculate the payback period, divide the total cost of the solar system by the annual savings on electricity bills. This calculation provides a clear picture of how long it will take for the system to pay for itself.

Several factors can influence the payback period, including the efficiency of the solar panels, the cost of electricity, and the availability of incentives. High-efficiency panels can generate more electricity, leading to greater savings and a shorter payback period. Similarly, in areas with high electricity costs, the savings from solar energy can be more substantial, reducing the payback period. Additionally, government incentives can significantly shorten the payback period by reducing the initial cost of the system.

It's also important to consider the lifespan of the solar panels when calculating the ROI and payback period. Most solar panels come with a warranty of 25 to 30 years, during which they are expected to operate at a high level of efficiency. This long lifespan means that even after the payback period is reached, homeowners can continue to enjoy savings on their electricity bills for many years. The durability and longevity of solar panels make them a sound investment for those looking to reduce their carbon footprint and save money in the long run.

Case studies of homeowners who have successfully installed solar energy systems can provide valuable insights into the ROI and payback period. These real-world examples illustrate the financial benefits of solar energy and highlight the factors that contribute to a successful investment. By examining these case studies, homeowners can gain a better understanding of what to expect and how to maximize the benefits of their solar energy system.

In addition to the financial benefits, investing in solar energy also offers environmental advantages. By reducing reliance on fossil fuels, solar energy helps decrease greenhouse gas emissions and combat climate change. This environmental impact can be an important consideration for homeowners who are committed to sustainability and want to make a positive contribution to the planet.

When evaluating the ROI and payback period, it's essential to consider the potential for future increases in electricity rates. As the cost of electricity continues to rise, the savings from solar energy are likely to increase, further enhancing the ROI and shortening the payback period. This potential for increased savings makes solar energy an even more attractive investment for homeowners looking to protect themselves from rising energy costs.

It's also worth noting that the value of a home can increase with the installation of a solar energy system. Many homebuyers are attracted to properties with solar panels, as they offer the promise of lower energy bills and a reduced carbon footprint. This increased demand can lead to higher property values, providing an additional financial benefit to homeowners who invest in solar energy.

In conclusion, calculating the ROI and payback period for a solar energy system is a critical step in assessing its financial viability. By considering factors such as initial costs, energy savings, government incentives, and the lifespan of the panels, homeowners can make informed decisions about the benefits of solar energy. With the potential for significant long-term savings, environmental benefits, and increased property values, solar energy represents a compelling investment for those looking to embrace renewable energy and reduce their reliance on traditional power sources.

Case Studies: Financial Success Stories

In the realm of solar energy, financial success stories abound, each one a testament to the transformative power of harnessing the sun's energy. These case studies not only illustrate the economic benefits of solar power but also highlight the diverse ways in which individuals and communities have leveraged this renewable resource to achieve financial independence and sustainability.

Consider the story of a small suburban family in California, who decided to invest in solar panels after years of grappling with soaring electricity bills. With an initial investment of $15,000, they installed a 5-kilowatt solar system on their rooftop. The family took advantage of federal tax credits and state rebates, which significantly reduced their upfront costs. Within the first year, they noticed a dramatic decrease in their electricity bills, saving approximately $1,500 annually. Over the next five years, these savings compounded, allowing them to recoup their initial investment. Today, they enjoy virtually free electricity, and their home has increased in value due to the solar installation.

In another instance, a rural farming community in the Midwest banded together to create a solar cooperative. Faced with the dual challenges of high energy costs and the need for sustainable farming practices, the community pooled their resources to install a large solar array. By doing so, they not only reduced their collective energy expenses but also generated additional income by selling excess power back to the grid. This initiative not only improved the financial stability of the farmers but also fostered a sense of community and shared purpose. The cooperative model has since been replicated in neighboring areas, demonstrating the scalability and impact of community-driven solar projects.

A tech-savvy entrepreneur in Texas saw an opportunity to integrate solar energy with smart home technology. By installing solar panels and pairing them with a state-of-the-art energy management system, he was able to optimize his energy consumption and maximize savings. The system allowed him to monitor energy usage in real-time, adjust settings remotely, and store excess energy in a battery for later use. This innovative approach not only reduced his energy bills by 70% but also provided a blueprint for others looking to combine solar power with modern technology. His success story has inspired a wave of tech enthusiasts to explore similar integrations, further driving the adoption of solar energy.

In urban settings, solar energy has also proven to be a game-changer for businesses. A boutique hotel in New York City decided to go green by installing solar panels on its rooftop. The hotel's management was initially motivated by the desire to reduce its carbon footprint and appeal to environmentally conscious travelers. However, the financial benefits quickly became apparent. The solar installation led to a 40% reduction in energy costs, allowing the hotel to reinvest savings into enhancing guest experiences. Additionally, the hotel's commitment to sustainability attracted a new clientele, boosting occupancy rates and revenue. This success story underscores the potential for businesses to achieve both environmental and financial gains through solar energy.

Educational institutions have also embraced solar power as a means of reducing operational costs and promoting sustainability. A public school district in Arizona embarked on a solar initiative to offset rising energy expenses and allocate more funds to educational programs. By installing solar panels across multiple school buildings, the district achieved substantial energy savings, which were redirected towards improving classroom resources and facilities. The project also served as an educational tool, providing students with hands-on learning opportunities about renewable energy and environmental stewardship. This case study highlights the multifaceted benefits of solar energy in educational settings, where financial savings can directly enhance the quality of education.

In the realm of residential real estate, solar energy has become a valuable asset for homeowners looking to increase property value and marketability. A couple in Florida decided to install solar panels before putting their home on the market. The investment paid off handsomely, as the solar-equipped home attracted multiple offers and sold at a premium compared to similar properties without solar installations. This trend is becoming increasingly common, as homebuyers recognize the long-term savings and environmental benefits of solar energy. Real estate agents are now touting solar installations as a key selling point, further driving demand for solar-equipped homes.

These financial success stories illustrate the diverse ways in which solar energy can be harnessed to achieve economic benefits. Whether through individual investments, community initiatives, or business ventures, solar power offers a pathway to financial independence and sustainability. The common thread in these stories is the strategic use of available incentives, innovative approaches to energy management, and a commitment to long-term savings. As more individuals and organizations embrace solar energy, the potential for financial success continues to grow, paving the way for a brighter, more sustainable future.

Chapter 4: Choosing the Right Solar System

Comparing Residential Solar Systems

Choosing the right solar system for your home can be a daunting task, given the variety of options available in the market. Each type of solar system comes with its own set of advantages and considerations, making it essential to understand the differences to make an informed decision. The primary types of residential solar systems include grid-tied, off-grid, and hybrid systems, each catering to different energy needs and preferences.

Grid-tied solar systems are the most common type of residential solar installation. These systems are connected to the local utility grid, allowing homeowners to draw electricity from the grid when their solar panels are not producing enough power, such as during nighttime or cloudy days. One of the main benefits of grid-tied systems is the ability to participate in net metering programs. Net metering allows homeowners to earn credits for the excess electricity their solar panels generate and send back to the grid. These credits can offset the cost of electricity drawn from the grid, leading to significant savings on utility bills. Grid-tied systems are generally more affordable than other types, as they do not require expensive battery storage. However, they do rely on the grid for backup power, which means they are not immune to power outages.

Off-grid solar systems, on the other hand, are completely independent of the utility grid. These systems are ideal for remote locations where grid access is limited or unavailable. Off-grid systems require battery storage to store excess energy generated during the day for use at night or during periods of low sunlight. While off-grid systems offer complete energy independence, they come with higher upfront costs due to the need for batteries and additional equipment. Homeowners must carefully size their system to ensure it can meet their energy needs year-round, taking into account seasonal variations in sunlight. Off-grid systems also require more maintenance and monitoring to ensure optimal performance.

Hybrid solar systems combine the best of both grid-tied and off-grid systems. These systems are connected to the grid but also include battery storage to provide backup power during outages. Hybrid systems offer greater flexibility and energy security, allowing homeowners to store excess solar energy for later use and reduce their reliance on the grid. This can be particularly beneficial in areas prone to power outages or with high electricity rates. While hybrid systems are more expensive than grid-tied systems, they offer the advantage of energy independence without the need to go completely off-grid.

When comparing residential solar systems, it's important to consider factors such as cost, energy needs, location, and long-term goals. The initial cost of a solar system can vary widely depending on the type of system, the size of the installation, and the quality of the components. Grid-tied systems are typically the most cost-effective option, while off-grid and hybrid systems require a larger investment due to the need for batteries and additional equipment. However, the long-term savings and benefits of each system should also be taken into account.

Energy needs play a crucial role in determining the most suitable solar system for a home. Homeowners should assess their average energy consumption and peak usage times to ensure their chosen system can meet their needs. For those with high energy demands or who wish to achieve complete energy independence, an off-grid or hybrid system may be more appropriate. Conversely, homeowners with moderate energy needs and access to net metering programs may find a grid-tied system to be the most cost-effective solution.

Location is another important consideration when comparing solar systems. The amount of sunlight a location receives can impact the efficiency and performance of a solar system. Areas with abundant sunlight may benefit more from solar installations, while regions with less consistent sunlight may require larger systems or additional components to achieve the desired energy output. Additionally, local regulations and incentives can influence the choice of solar system, as some areas offer more favorable conditions for certain types of installations.

Long-term goals and personal preferences should also be factored into the decision-making process. Homeowners who prioritize energy independence and sustainability may be more inclined to invest in off-grid or hybrid systems, despite the higher upfront costs. Those who are primarily motivated by financial savings may opt for a grid-tied system to take advantage of net metering and lower installation costs. It's important to weigh the pros and cons of each system type in relation to individual goals and circumstances.

In addition to the type of solar system, homeowners should also consider the quality and efficiency of the solar panels and inverters used in the installation. High-quality components can improve the overall performance and longevity of the system, leading to greater savings and a higher return on investment. It's advisable to research and compare different brands and models to ensure the chosen system meets the desired standards of efficiency and reliability.

Working with a reputable solar installer can also make a significant difference in the success of a solar project. Experienced installers can provide valuable insights and recommendations based on local conditions and individual needs. They can also assist with navigating the complexities of permits, incentives, and financing options, ensuring a smooth and hassle-free installation process.

Ultimately, the choice of a residential solar system depends on a variety of factors, including budget, energy needs, location, and personal preferences. By carefully considering these factors and comparing the different types of systems available, homeowners can make an informed decision that aligns with their goals and maximizes the benefits of solar energy. Whether opting for a grid-tied, off-grid, or hybrid system, the transition to solar power represents a significant step towards energy independence and sustainability.

Selecting the Best Solar Panels for Your Needs

Selecting the best solar panels for your needs involves a careful evaluation of various factors that can influence the efficiency, cost, and overall performance of your solar energy system. With the growing popularity of solar power, the market offers a wide array of options, each with its own set of characteristics and benefits. Understanding these differences is crucial to making an informed decision that aligns with your energy goals and budget.

The first consideration when choosing solar panels is the type of photovoltaic technology used. The three main types of solar panels are monocrystalline, polycrystalline, and thin-film. Monocrystalline panels are known for their high efficiency and sleek appearance. They are made from a single crystal structure, which allows electrons to move more freely, resulting in higher energy conversion rates. These panels are ideal for homes with limited roof space, as they generate more power per square foot. However, they tend to be more expensive than other types.

Polycrystalline panels, on the other hand, are made from multiple silicon crystals melted together. While they are generally less efficient than monocrystalline panels, they are also more affordable, making them a popular choice for homeowners looking to balance cost and performance. Polycrystalline panels have a bluish hue and a slightly lower efficiency, but they can still provide substantial energy savings, especially in areas with ample sunlight.

Thin-film solar panels are the most versatile in terms of application. They are made by depositing one or more layers of photovoltaic material onto a substrate, such as glass or metal. Thin-film panels are lightweight and flexible, making them suitable for a variety of installations, including curved surfaces and portable applications. While they are generally less efficient than crystalline panels, their lower cost and adaptability make them an attractive option for certain projects.

Efficiency is a key factor to consider when selecting solar panels. The efficiency of a solar panel refers to the percentage of sunlight that is converted into usable electricity. Higher efficiency panels can generate more power in a given area, which is particularly important for homes with limited roof space. However, higher efficiency panels often come with a higher price tag, so it's important to weigh the benefits against the cost.

Durability and warranty are also important considerations. Solar panels are a long-term investment, and their performance over time can significantly impact the return on investment. Most solar panels come with a warranty that guarantees a certain level of performance for a specified period, typically 25 years. It's important to choose panels from reputable manufacturers that offer strong warranties and have a track record of reliability.

The aesthetic appeal of solar panels can also influence your decision, especially if you are concerned about the visual impact on your home. Monocrystalline panels are often favored for their uniform, black appearance, which can blend more seamlessly with certain roof types. Polycrystalline panels have a more speckled blue appearance, which may be more noticeable. Thin-film panels offer the most flexibility in terms of design, as they can be integrated into building materials for a more discreet look.

Cost is a major factor for most homeowners when selecting solar panels. While it's tempting to opt for the cheapest option, it's important to consider the long-term savings and benefits of higher quality panels. Investing in more efficient and durable panels can lead to greater energy savings and a higher return on investment over time. Additionally, government incentives and rebates can help offset the initial cost, making higher quality panels more accessible.

The installation process and compatibility with your existing infrastructure should also be considered. Some panels may require specific mounting systems or inverters, which can affect the overall cost and complexity of the installation. It's important to work with a qualified solar installer who can assess your home's specific needs and recommend the best panels for your situation.

Environmental impact is another consideration for eco-conscious homeowners. While all solar panels contribute to reducing carbon emissions, the manufacturing process and materials used can vary in terms of environmental friendliness. Some manufacturers prioritize sustainable practices and use recycled or low-impact materials, which can further enhance the environmental benefits of your solar system.

Finally, it's important to consider your long-term energy goals and how solar panels fit into your overall energy strategy. Whether you are looking to reduce your carbon footprint, achieve energy independence, or simply save money on electricity bills, selecting the right solar panels is a crucial step in achieving those goals. By carefully evaluating the various options and considering factors such as efficiency, cost, durability, and aesthetics, you can make an informed decision that meets your needs and maximizes the benefits of solar energy.

In conclusion, selecting the best solar panels for your needs requires a comprehensive evaluation of various factors, including photovoltaic technology, efficiency, durability, cost, and aesthetics. By understanding the differences between monocrystalline, polycrystalline, and thin-film panels, and considering your specific energy goals and budget, you can make an informed decision that aligns with your needs and maximizes the benefits of solar energy. With the right panels, you can enjoy significant energy savings, reduce your carbon footprint, and contribute to a more sustainable future.

Understanding Warranties and Maintenance Requirements

Navigating the world of solar energy involves understanding not only the technology and financial aspects but also the warranties and maintenance requirements that come with your investment. These elements are crucial in ensuring the longevity and efficiency of your solar system, providing peace of mind and protecting your investment over the long term.

Warranties are a key component of any solar panel purchase, offering protection against defects and performance issues. Typically, solar panels come with two types of warranties: product (or equipment) warranties and performance warranties. Product warranties cover defects in materials and workmanship, ensuring that the panels are free from manufacturing flaws. These warranties usually last between 10 to 25 years, depending on the manufacturer. A longer product warranty is often indicative of the manufacturer's confidence in the durability and quality of their panels.

Performance warranties, on the other hand, guarantee that the solar panels will maintain a certain level of energy output over time. Solar panels naturally degrade as they age, resulting in a gradual decrease in efficiency. Performance warranties typically promise that the panels will retain a specific percentage of their original efficiency after a set number of years, often 80-90% after 25 years. This type of warranty is crucial for ensuring that your solar system continues to meet your energy needs well into the future.

When evaluating warranties, it's important to consider the reputation and reliability of the manufacturer. Established manufacturers with a proven track record are more likely to honor their warranties and provide support if issues arise. It's also worth noting that some manufacturers offer extended warranties for an additional cost, which can provide extra protection and peace of mind.

In addition to panel warranties, it's essential to consider the warranties for other components of your solar system, such as inverters and mounting equipment. Inverters, which convert the direct current (DC) generated by the panels into alternating current (AC) for use in your home, typically have shorter warranties than the panels themselves, often ranging from 5 to 15 years. Given the critical role inverters play in the overall system, it's advisable to choose high-quality inverters with robust warranties.

Maintenance requirements for solar systems are generally minimal, but regular upkeep is essential to ensure optimal performance and longevity. One of the primary maintenance tasks is keeping the solar panels clean and free from debris. Dust, dirt, leaves, and bird droppings can accumulate on the panels, reducing their efficiency by blocking sunlight. In most cases, rain will naturally clean the panels, but in dry or dusty areas, periodic cleaning may be necessary. This can be done with a hose or a soft brush and mild detergent, taking care not to scratch the surface of the panels.

It's also important to regularly inspect the panels and mounting equipment for any signs of damage or wear. Look for cracks, chips, or discoloration on the panels, as well as loose or corroded mounting hardware. Addressing these issues promptly can prevent more significant problems and ensure the system continues to operate efficiently.

Monitoring the performance of your solar system is another crucial aspect of maintenance. Many modern solar systems come with monitoring software that allows you to track energy production and consumption in real-time. This can help you identify any drops in performance that may indicate a problem with the system. If you notice a significant decrease in energy output, it's important to investigate the cause and address any issues promptly.

Inverters, being a critical component of the solar system, may require occasional maintenance or replacement. While they are generally reliable, inverters can experience issues over time, such as overheating or electrical faults. Regularly checking the inverter's display for error messages and ensuring it is operating within the recommended temperature range can help prevent problems. If an inverter does fail, it may need to be repaired or replaced, which is why having a good warranty is important.

Working with a reputable solar installer can also simplify the maintenance process. Many installers offer maintenance packages or service agreements that include regular inspections, cleaning, and performance monitoring. These services can provide peace of mind and ensure that your solar system continues to operate at peak efficiency.

Understanding the warranties and maintenance requirements of your solar system is essential for protecting your investment and ensuring long-term performance. By choosing high-quality components with robust warranties and committing to regular maintenance, you can maximize the benefits of solar energy and enjoy reliable, sustainable power for years to come. Whether you're a new solar adopter or a seasoned enthusiast, taking the time to understand these aspects will help you make the most of your solar energy system.

Working with Solar Installers: What to Expect

Embarking on the journey to solar energy begins with selecting the right solar installer, a crucial step that can significantly impact the success and efficiency of your solar project. Understanding what to expect when working with solar installers can help you navigate the process smoothly and ensure that your investment yields the desired results. From initial consultations to post-installation support, each phase of the process requires careful consideration and communication.

The first step in working with a solar installer is the initial consultation. During this phase, the installer will assess your energy needs, evaluate your property, and discuss your goals for the solar system. This is an opportunity for you to ask questions and gain a clear understanding of the potential benefits and limitations of solar energy for your home. A reputable installer will provide a detailed analysis of your current energy consumption, the potential energy savings, and the estimated cost of the solar system. They should also explain the different types of solar panels and inverters available, helping you make an informed decision based on your specific needs and budget.

Once the initial consultation is complete, the installer will conduct a site assessment to evaluate the suitability of your property for solar installation. This involves examining factors such as roof orientation, shading, and structural integrity. The installer will also assess the electrical system to ensure it can accommodate the new solar equipment. Based on this assessment, the installer will design a customized solar system tailored to your property's unique characteristics. This design will include the layout of the solar panels, the type of inverters to be used, and any additional equipment required for optimal performance.

After the site assessment and system design, the installer will provide a detailed proposal outlining the scope of work, the estimated timeline, and the total cost of the project. This proposal should include a breakdown of the costs for equipment, installation, permits, and any additional services. It's important to review this proposal carefully and ensure that all aspects of the project are clearly defined. If you have any questions or concerns, don't hesitate to discuss them with the installer before proceeding.

Once you have agreed to the proposal, the installer will handle the necessary permits and approvals required for the solar installation. This process can vary depending on local regulations and utility requirements, but a reputable installer will have experience navigating these complexities and will ensure that all necessary paperwork is completed accurately and promptly. This step is crucial to avoid any delays or complications during the installation process.

The installation phase is where the solar system is physically installed on your property. This process typically takes a few days to a week, depending on the size and complexity of the system. The installer will coordinate the delivery of equipment, prepare the site, and install the solar panels, inverters, and any additional components. They will also connect the system to your electrical panel and ensure that it is functioning correctly. Throughout the installation, the installer should keep you informed of the progress and address any questions or concerns you may have.

Once the installation is complete, the system will undergo a series of inspections and tests to ensure it meets all safety and performance standards. This includes inspections by local authorities and utility companies, as well as testing by the installer to verify that the system is operating as expected. Once all inspections are passed, the system will be activated, and you can begin enjoying the benefits of solar energy.

Post-installation support is an important aspect of working with a solar installer. A reputable installer will provide ongoing support and maintenance services to ensure your solar system continues to operate efficiently. This may include regular inspections, cleaning, and performance monitoring. Additionally, the installer should be available to address any issues or concerns that may arise after the installation. It's important to establish a clear understanding of the support services offered and any associated costs.

When selecting a solar installer, it's important to consider factors such as experience, reputation, and customer reviews. Look for installers with a proven track record of successful installations and satisfied customers. It's also advisable to obtain multiple quotes from different installers to compare pricing, services, and warranties. This will help you make an informed decision and ensure you receive the best value for your investment.

Communication is key when working with a solar installer. Establishing clear lines of communication from the outset will help ensure that the project runs smoothly and any issues are addressed promptly. Be sure to discuss your expectations, ask questions, and provide feedback throughout the process. A good installer will value your input and work collaboratively to achieve the best possible outcome.

In summary, working with a solar installer involves several key steps, from initial consultations and site assessments to installation and post-installation support. By understanding what to expect and actively participating in the process, you can ensure a successful solar installation that meets your energy needs and provides long-term benefits. With the right installer, you can confidently transition to solar energy and enjoy the many advantages it offers.

DIY Solar Installation: Pros and Cons

Embarking on a DIY solar installation can be an enticing prospect for those who relish the challenge of hands-on projects and seek to save on installation costs. However, this path is not without its complexities and potential pitfalls. Understanding the pros and cons of a DIY solar installation is crucial for making an informed decision that aligns with your skills, resources, and energy goals.

One of the most compelling advantages of a DIY solar installation is the potential for cost savings. By eliminating the labor costs associated with hiring professional installers, homeowners can significantly reduce the overall expense of their solar project. This can make solar energy more accessible for those on a tight budget, allowing them to invest in higher-quality panels or additional components. Additionally, some DIY enthusiasts find satisfaction in the process itself, gaining a sense of accomplishment from completing the installation independently.

Another benefit of a DIY approach is the flexibility it offers. Homeowners can work at their own pace, taking the time to research and select the best components for their specific needs. This can lead to a more customized solar system that is tailored to the unique characteristics of the property and the homeowner's energy consumption patterns. DIY installers also have the freedom to choose from a wider range of products, as they are not limited to the brands and models offered by professional installers.

However, the DIY route is not without its challenges. One of the primary drawbacks is the complexity of the installation process. Installing a solar system requires a solid understanding of electrical systems, roofing, and local building codes. Mistakes during installation can lead to safety hazards, reduced system performance, or even damage to the property. For those without prior experience or technical expertise, the learning curve can be steep and time-consuming.

Permitting and regulatory compliance present another hurdle for DIY installers. Navigating the permitting process can be daunting, as it involves understanding and adhering to local regulations and utility requirements. Failure to obtain the necessary permits or comply with building codes can result in fines, delays, or the need to dismantle and redo the installation. Professional installers typically handle these aspects on behalf of their clients, ensuring that all legal requirements are met.

Safety is a critical consideration in any solar installation, and DIY projects are no exception. Working on rooftops and with electrical systems poses inherent risks, including falls, electrical shocks, and fire hazards. It's essential for DIY installers to prioritize safety by using appropriate protective gear, following best practices, and adhering to all safety guidelines. For those who are not comfortable with these risks, hiring a professional installer may be a safer option.

Another potential downside of a DIY solar installation is the lack of warranties and support. Many solar panel manufacturers offer warranties that are contingent on professional installation. By opting for a DIY approach, homeowners may forfeit these warranties, leaving them without recourse if issues arise. Additionally, professional installers often provide ongoing support and maintenance services, which can be invaluable for ensuring the long-term performance of the system. DIY installers must be prepared to troubleshoot and resolve any problems independently.

Despite these challenges, a successful DIY solar installation is achievable with careful planning, research, and preparation. Homeowners considering this route should start by thoroughly educating themselves on the components and processes involved. Numerous online resources, forums, and instructional videos are available to guide DIY enthusiasts through each step of the installation. It's also advisable to consult with professionals or experienced DIYers to gain insights and advice.

Before embarking on a DIY solar project, it's important to assess your skills, resources, and commitment to the task. Consider whether you have the necessary tools, time, and expertise to complete the installation safely and effectively. If you have any doubts about your ability to handle the project, it may be worth exploring hybrid options, such as hiring a professional for certain aspects of the installation while completing others yourself.

In conclusion, a DIY solar installation offers both opportunities and challenges. While the potential for cost savings and customization is appealing, the complexities and risks involved require careful consideration. By weighing the pros and cons and thoroughly preparing for the task, homeowners can make an informed decision that aligns with their goals and capabilities. Whether choosing to go it alone or enlisting professional help, the transition to solar energy represents a significant step towards sustainability and energy independence.

Chapter 5: Maximizing Solar Efficiency

Optimizing Panel Placement and Angle

Harnessing the full potential of solar energy hinges on the strategic placement and angling of solar panels. This aspect of solar installation is crucial, as it directly influences the efficiency and energy output of the system. Understanding how to optimize panel placement and angle can significantly enhance the performance of your solar array, ensuring you get the most out of your investment.

The first consideration in optimizing solar panel placement is the orientation of the panels. In the northern hemisphere, panels should ideally face true south to capture the maximum amount of sunlight throughout the day. Conversely, in the southern hemisphere, panels should face true north. This orientation allows the panels to receive the most direct sunlight, maximizing energy production. However, due to the magnetic declination, which is the angle difference between magnetic north and true north, it's important to adjust the orientation slightly to account for this variation. Tools like a compass or a smartphone app can help determine the correct orientation for your location.

Roof pitch and angle play a significant role in optimizing solar panel performance. The angle of the panels should be adjusted to match the latitude of your location, as this allows the panels to capture the most sunlight throughout the year. For instance, if you live at a latitude of 30 degrees, the optimal angle for your panels would be approximately 30 degrees. This angle can be adjusted seasonally to account for the sun's changing position in the sky, although fixed angles are more common for residential installations due to their simplicity and lower cost.

Shading is another critical factor to consider when optimizing panel placement. Even partial shading on a single panel can significantly reduce the overall efficiency of the entire system. It's essential to conduct a thorough shading analysis to identify any potential obstructions, such as trees, chimneys, or neighboring buildings, that could cast shadows on the panels. If shading is unavoidable, microinverters or power optimizers can be used to mitigate the impact by allowing each panel to operate independently.

The structural integrity of the roof is also a key consideration. Solar panels add weight to the roof, and it's important to ensure that the roof can support this additional load. A professional structural assessment can help determine if any reinforcements are needed before installation. Additionally, the condition of the roofing material should be evaluated, as installing solar panels on a roof that requires repair or replacement can lead to complications and additional costs down the line.

For those with limited roof space or unsuitable roof conditions, ground-mounted solar systems offer an alternative solution. These systems provide greater flexibility in terms of placement and angle, as they can be installed in open areas with optimal sun exposure. Ground-mounted systems can also be adjusted more easily to track the sun's movement, further enhancing energy production. However, they require additional space and may involve higher installation costs due to the need for mounting structures and trenching for electrical connections.

In urban environments where roof space is limited, innovative solutions such as solar canopies or solar awnings can be employed. These installations not only provide shade and shelter but also generate electricity, making efficient use of available space. Additionally, bifacial solar panels, which capture sunlight on both sides, can be used in areas with reflective surfaces to increase energy production.

Monitoring and maintenance are essential for ensuring that solar panels continue to operate at peak efficiency. Regularly cleaning the panels to remove dust, dirt, and debris can prevent a reduction in energy output. It's also important to inspect the panels and mounting equipment for any signs of wear or damage, addressing any issues promptly to avoid more significant problems.

Advancements in solar technology, such as solar tracking systems, offer further opportunities for optimizing panel placement and angle. These systems automatically adjust the position of the panels to follow the sun's movement, maximizing energy capture throughout the day. While solar tracking systems can increase energy production by up to 25-35%, they also involve higher costs and maintenance requirements, making them more suitable for larger installations or commercial projects.

Incorporating energy storage solutions, such as batteries, can complement the optimization of panel placement and angle by storing excess energy generated during peak sunlight hours for use during periods of low sunlight or at night. This can enhance the overall efficiency and reliability of the solar system, providing greater energy independence and resilience.

Ultimately, optimizing solar panel placement and angle requires a comprehensive understanding of various factors, including orientation, angle, shading, and structural considerations. By carefully evaluating these elements and employing innovative solutions where necessary, homeowners can maximize the performance and benefits of their solar energy system. Whether through traditional roof-mounted panels or alternative installations, the strategic placement and angling of solar panels are key to unlocking the full potential of solar energy.

The Importance of Regular Maintenance

Regular maintenance is the unsung hero of solar energy systems, ensuring that they continue to operate efficiently and deliver the expected energy savings over their lifespan. While solar panels are known for their durability and low maintenance requirements, neglecting routine upkeep can lead to decreased performance and costly repairs. Understanding the importance of regular maintenance and implementing a proactive approach can maximize the benefits of your solar investment.

One of the primary reasons for regular maintenance is to maintain the efficiency of the solar panels. Over time, dust, dirt, leaves, and other debris can accumulate on the surface of the panels, obstructing sunlight and reducing their energy output. In areas prone to pollution or frequent dust storms, this buildup can be significant. Regular cleaning of the panels is essential to ensure they capture as much sunlight as possible. Depending on the location and environmental conditions, cleaning may be required every few months or more frequently. Using a soft brush or a hose with mild detergent can effectively remove debris without damaging the panels.

In addition to cleaning, regular inspections are crucial for identifying potential issues before they escalate into major problems. Inspecting the panels for cracks, chips, or discoloration can help detect damage that may affect performance. Similarly, checking the mounting hardware for signs of corrosion or looseness can prevent structural failures. These inspections should also extend to the electrical components of the system, including the inverters and wiring. Look for any signs of wear, overheating, or electrical faults, and address them promptly to avoid disruptions in energy production.

Monitoring the performance of the solar system is another key aspect of regular maintenance. Many modern solar systems come equipped with monitoring software that provides real-time data on energy production and consumption. By keeping an eye on this data, you can quickly identify any drops in performance that may indicate a problem. For instance, if the system is producing less energy than expected, it could be due to shading, dirt accumulation, or a malfunctioning component. Early detection allows for timely intervention, minimizing downtime and ensuring the system continues to operate at peak efficiency.

Inverters, which convert the direct current (DC) generated by the panels into alternating current (AC) for use in your home, are a critical component of the solar system. They are also one of the most common points of failure. Regularly checking the inverter's display for error messages and ensuring it operates within the recommended temperature range can help prevent issues. Inverters typically have a shorter lifespan than the panels themselves, so it's important to be aware of their warranty period and plan for eventual replacement.

While solar panels are designed to withstand harsh weather conditions, extreme events such as hailstorms, hurricanes, or heavy snowfall can cause damage. After such events, it's important to inspect the panels and mounting structures for any signs of impact or stress. Addressing any damage promptly can prevent further deterioration and ensure the system remains safe and functional.

Regular maintenance also plays a role in preserving the warranties that come with your solar system. Many manufacturers require proof of regular maintenance to honor their warranties. Keeping detailed records of cleaning, inspections, and any repairs or replacements can provide the necessary documentation if a warranty claim is needed. This not only protects your investment but also ensures that you receive the full benefits of the warranties provided.

For those who prefer a hands-off approach, many solar installers offer maintenance packages or service agreements. These services typically include regular inspections, cleaning, and performance monitoring, providing peace of mind and ensuring that the system continues to operate efficiently. While these services come at an additional cost, they can be a worthwhile investment for those who lack the time or expertise to perform maintenance themselves.

Incorporating energy storage solutions, such as batteries, into your solar system adds another layer of maintenance considerations. Batteries require regular checks to ensure they are operating within the recommended parameters and to prevent issues such as overcharging or deep discharging. Proper maintenance of batteries can extend their lifespan and enhance the overall reliability of the solar system.

Ultimately, regular maintenance is essential for maximizing the performance and longevity of your solar energy system. By adopting a proactive approach to cleaning, inspections, and monitoring, you can ensure that your system continues to deliver the expected energy savings and environmental benefits. Whether you choose to perform maintenance yourself or enlist professional services, the effort invested in regular upkeep will pay dividends in the form of a reliable and efficient solar energy system.

Monitoring System Performance

Monitoring the performance of a solar energy system is a critical aspect of ensuring its efficiency and longevity. By keeping a close eye on how well your system is functioning, you can maximize energy production, identify potential issues early, and make informed decisions about maintenance and upgrades. Understanding the tools and techniques available for monitoring can empower you to take full control of your solar investment.

At the heart of effective monitoring is the ability to track energy production and consumption in real-time. Many modern solar systems come equipped with monitoring software that provides detailed insights into the system's performance. This software can be accessed through a computer or smartphone app, offering a convenient way to stay informed about your system's output. By analyzing this data, you can determine whether your system is meeting its expected performance levels and identify any discrepancies that may indicate a problem.

One of the key metrics to monitor is the energy output of the solar panels. This is typically measured in kilowatt-hours (kWh) and can vary based on factors such as weather conditions, shading, and the time of year. By comparing the actual energy output to the expected output, you can assess whether the system is operating efficiently. If you notice a significant drop in energy production, it may be due to issues such as dirt accumulation on the panels, shading from nearby trees or buildings, or a malfunctioning component.

In addition to energy output, it's important to monitor the system's efficiency. This refers to the ratio of energy produced to the energy that could potentially be generated under ideal conditions. Efficiency can be affected by factors such as panel degradation, inverter performance, and temperature fluctuations. Regularly checking the efficiency of your system can help you identify areas for improvement and ensure that you are getting the most out of your solar investment.

Inverters play a crucial role in the overall performance of a solar system, as they convert the direct current (DC) generated by the panels into alternating current (AC) for use in your home. Monitoring the performance of the inverter is essential, as any issues with this component can significantly impact the system's output. Many inverters come with built-in monitoring capabilities, allowing you to track their performance and receive alerts if any faults are detected. Keeping an eye on the inverter's status and addressing any error messages promptly can prevent disruptions in energy production.

Temperature is another factor that can influence the performance of a solar system. Solar panels are most efficient at lower temperatures, and excessive heat can reduce their output. Monitoring the temperature of the panels and the surrounding environment can help you understand how temperature fluctuations affect your system's performance. In some cases, installing cooling systems or optimizing ventilation can help mitigate the impact of high temperatures and improve efficiency.

For those who want to take monitoring to the next level, advanced monitoring systems offer additional features such as weather forecasting, energy storage management, and detailed analytics. These systems can provide insights into how weather patterns affect energy production and help you plan for periods of low sunlight. They can also optimize the use of energy storage solutions, such as batteries, by predicting energy demand and adjusting the charging and discharging cycles accordingly.

Regularly reviewing the data collected by your monitoring system can help you identify trends and patterns in your energy production and consumption. This information can be invaluable for making informed decisions about system upgrades, maintenance, and energy usage. For example, if you notice that your energy consumption consistently exceeds production during certain times of the year, you may consider adding more panels or investing in energy storage solutions to meet your needs.

In addition to monitoring the performance of the solar system itself, it's important to keep an eye on your overall energy consumption. By understanding how and when you use energy, you can identify opportunities for energy conservation and efficiency improvements. Simple changes, such as adjusting your thermostat, using energy-efficient appliances, and reducing standby power consumption, can complement your solar system and further reduce your energy bills.

While monitoring systems provide valuable insights, it's important to remember that they are only as effective as the actions you take based on the data they provide. Regularly reviewing and analyzing the information collected by your monitoring system can help you make proactive decisions that enhance the performance and longevity of your solar energy system. Whether it's scheduling maintenance, optimizing energy usage, or planning for future upgrades, the insights gained from monitoring can guide you in making the most of your solar investment.

In conclusion, monitoring the performance of your solar energy system is an essential practice that can significantly impact its efficiency and lifespan. By leveraging the tools and techniques available for monitoring, you can gain a deeper understanding of your system's performance, identify potential issues early, and make informed decisions that maximize the benefits of solar energy. With a proactive approach to monitoring, you can ensure that your solar system continues to deliver reliable and sustainable energy for years to come.

Integrating Energy Storage Solutions

Integrating energy storage solutions into a solar energy system is a transformative step that can significantly enhance the efficiency, reliability, and flexibility of your energy usage. As solar power generation is inherently intermittent, relying solely on solar panels can lead to periods of surplus energy during sunny days and shortages during cloudy days or at night. Energy storage solutions, such as batteries, bridge this gap by storing excess energy for use when solar production is low, providing a more consistent and dependable power supply.

The most common form of energy storage for residential solar systems is the lithium-ion battery. Known for their high energy density, long lifespan, and declining costs, lithium-ion batteries have become the go-to choice for homeowners looking to store solar energy. These batteries can be integrated seamlessly with solar panels and inverters, allowing for efficient energy management. When selecting a battery, it's important to consider factors such as capacity, power rating, depth of discharge, and round-trip efficiency. Capacity refers to the total amount of energy the battery can store, while power rating indicates how much energy can be delivered at once. Depth of discharge measures how much of the battery's capacity can be used without affecting its lifespan, and round-trip efficiency reflects the energy loss during the charging and discharging process.

Beyond lithium-ion, other battery technologies such as lead-acid, nickel-cadmium, and flow batteries are available, each with its own set of advantages and limitations. Lead-acid batteries, for instance, are more affordable but have a shorter lifespan and lower efficiency compared to lithium-ion. Flow batteries offer long cycle life and scalability but are typically more expensive and better suited for larger installations. Understanding the characteristics of each battery type can help you make an informed decision based on your specific needs and budget.

Integrating energy storage solutions involves more than just selecting the right battery. Proper installation and configuration are crucial for ensuring optimal performance and safety. Batteries should be installed in a well-ventilated area, away from extreme temperatures and moisture, to prevent overheating and degradation. Additionally, the battery management system (BMS) plays a vital role in monitoring and controlling the battery's state of charge, temperature, and overall health. A robust BMS can prevent overcharging, deep discharging, and other conditions that could damage the battery or reduce its lifespan.

One of the key benefits of energy storage is the ability to achieve greater energy independence. By storing excess solar energy, homeowners can reduce their reliance on the grid and minimize their exposure to fluctuating electricity prices. This is particularly advantageous in regions with time-of-use pricing, where electricity costs vary throughout the day. By using stored energy during peak pricing periods, homeowners can significantly lower their energy bills and increase their savings.

Energy storage also enhances the resilience of a solar energy system by providing backup power during grid outages. In the event of a power failure, a well-integrated battery system can supply electricity to essential appliances and devices, ensuring that your home remains functional and comfortable. This capability is especially valuable in areas prone to natural disasters or frequent power interruptions, offering peace of mind and security.

For those looking to maximize the benefits of energy storage, integrating smart energy management systems can further optimize energy usage. These systems use advanced algorithms and real-time data to predict energy demand, solar production, and battery state of charge. By intelligently managing the charging and discharging cycles, smart energy management systems can enhance the efficiency and lifespan of the battery while ensuring that energy is used in the most cost-effective manner.

In addition to residential applications, energy storage solutions are increasingly being adopted in commercial and industrial settings. Businesses can leverage energy storage to manage demand charges, which are fees based on the highest level of power drawn from the grid during a billing period. By using stored energy to reduce peak demand, businesses can lower their electricity costs and improve their bottom line. Furthermore, energy storage can support the integration of renewable energy sources, such as wind and solar, into the grid, contributing to a more sustainable and resilient energy infrastructure.

As the energy storage market continues to evolve, new technologies and innovations are emerging that promise to further enhance the capabilities and affordability of storage solutions. Solid-state batteries, for example, offer the potential for higher energy density, faster charging times, and improved safety compared to traditional lithium-ion batteries. Similarly, advancements in battery recycling and second-life applications are addressing environmental concerns and extending the value of energy storage systems.

Integrating energy storage solutions into a solar energy system is a strategic investment that offers numerous benefits, from increased energy independence and cost savings to enhanced resilience and sustainability. By carefully selecting the right battery technology, ensuring proper installation and management, and leveraging smart energy management systems, homeowners and businesses can unlock the full potential of their solar energy systems. As the world moves towards a cleaner and more sustainable energy future, energy storage will play a pivotal role in enabling the widespread adoption of renewable energy and transforming the way we generate, store, and consume power.

Enhancing Efficiency with Smart Home Technology

The integration of smart home technology with solar energy systems is revolutionizing the way we manage and consume energy. By enhancing efficiency and optimizing energy usage, smart home devices offer homeowners the ability to maximize the benefits of their solar installations. This chapter delves into the various ways smart home technology can be leveraged to improve energy efficiency, reduce costs, and create a more sustainable living environment.

At the core of smart home technology is the concept of automation and connectivity. Smart devices communicate with each other and with the homeowner, providing real-time data and control over various aspects of the home. When integrated with a solar energy system, these devices can intelligently manage energy consumption, ensuring that solar power is used effectively and efficiently.

One of the most impactful smart home technologies is the smart thermostat. These devices learn the homeowner's schedule and preferences, adjusting heating and cooling settings to optimize comfort while minimizing energy use. By coordinating with the solar system, a smart thermostat can prioritize the use of solar energy during peak production times, reducing reliance on the grid and lowering electricity bills. Additionally, smart thermostats can be controlled remotely via smartphone apps, allowing homeowners to make adjustments on the go and respond to changing weather conditions.

Smart lighting systems offer another avenue for enhancing energy efficiency. These systems use sensors and automation to adjust lighting based on occupancy and natural light levels. By ensuring that lights are only on when needed, smart lighting can significantly reduce energy consumption. Furthermore, smart lighting can be programmed to align with solar production, using solar energy during the day and minimizing grid usage at night. The ability to control lighting remotely adds an extra layer of convenience and efficiency.

Appliances are another major area where smart home technology can make a difference. Smart appliances, such as refrigerators, washing machines, and dishwashers, can be programmed to operate during times of peak solar production, maximizing the use of solar energy and reducing grid dependency. Some smart appliances even offer energy-saving modes that adjust their operation to minimize power consumption without sacrificing performance. By integrating these appliances with a solar energy system, homeowners can achieve greater energy efficiency and cost savings.

Energy monitoring systems are a crucial component of a smart home, providing detailed insights into energy production and consumption. These systems track the performance of the solar panels, inverters, and other components, offering real-time data on energy usage patterns. By analyzing this data, homeowners can identify areas for improvement and make informed decisions about energy management. Energy monitoring systems can also alert homeowners to potential issues, such as a drop in solar production or an increase in energy consumption, allowing for timely intervention and maintenance.

Smart home technology extends beyond individual devices, encompassing entire home automation systems that integrate various components into a cohesive network. These systems use advanced algorithms and machine learning to optimize energy usage, taking into account factors such as weather forecasts, energy prices, and the homeowner's schedule. By intelligently managing the operation of smart devices, home automation systems can enhance the efficiency and effectiveness of a solar energy system, ensuring that energy is used in the most cost-effective and sustainable manner.

The benefits of integrating smart home technology with solar energy systems extend beyond energy efficiency and cost savings. By reducing energy consumption and reliance on the grid, smart home technology contributes to a more sustainable and environmentally friendly lifestyle. This aligns with the broader goals of reducing carbon emissions and promoting renewable energy sources, making smart home technology an essential component of a modern, eco-conscious home.

For those considering the integration of smart home technology, it's important to assess the compatibility of existing devices and systems. Many smart home devices are designed to work with specific platforms, such as Amazon Alexa, Google Assistant, or Apple HomeKit. Ensuring that devices are compatible with the chosen platform can simplify installation and operation, providing a seamless and user-friendly experience.

Security is another important consideration when implementing smart home technology. As these devices are connected to the internet, they can be vulnerable to cyber threats. Taking steps to secure the home network, such as using strong passwords, enabling two-factor authentication, and keeping software up to date, can help protect against unauthorized access and ensure the safety and privacy of the homeowner.

The integration of smart home technology with solar energy systems represents a significant advancement in the pursuit of energy efficiency and sustainability. By leveraging the capabilities of smart devices, homeowners can optimize energy usage, reduce costs, and contribute to a more sustainable future. As technology continues to evolve, the potential for further enhancements in energy efficiency and smart home integration will only grow, offering exciting opportunities for those looking to embrace a smarter, more sustainable way of living.

www.ingramcontent.com/pod-product-compliance
Lightning Source LLC
Chambersburg PA
CBHW072034150726
47999CB00002B/898